SMARTPHONES

NANCY DICKMANN

BROWN BEAR BOOKS

Published by Brown Bear Books Ltd
4877 N. Circulo Bujia, Tucson, AZ 85718
USA
and
Studio G14, Regent Studios, 1 Thane Villas,
London N7 7PH, UK

Text: Nancy Dickmann
Design and Illustrations: Square and Circus
Design Manager: Keith Davis
Children's Publisher: Anne O'Daly

Library of Congress Cataloging-in-Publication Data
Names: Dickmann, Nancy, author.
Title: Smartphones / by Nancy Dickmann.
Description: Tuscon, AZ, USA : Brown Bear Books Ltd, [2025] | Series: Science starters | Includes bibliographical references and index. | Audience: Ages 6-9 | Audience: Grades 2-3 | Summary: "Find out how smartphones developed, how they work, and how people use them"– Provided by publisher.
Identifiers: LCCN 2023052979 (print) | LCCN 2023052980 (ebook) | ISBN 9781781219928 (library binding) | ISBN 9781781219980 (paperback) | ISBN 9781835720042 (ebook)
Subjects: LCSH: Smartphones–Juvenile literature.
Classification: LCC TK6564.4.C45 D53 2025 (print) | LCC TK6564.4.C45 (ebook) | DDC 004.167–dc23/eng/20231120
LC record available at https://lccn.loc.gov/2023052979
LC ebook record available at https://lccn.loc gov/2023052980

Picture Credits
The photographs in this book are used by permission and through the courtesy of:
Interior: Shutterstock: AtStock Productions 16–17, Casezy idea 10–11, Stan Drett 14–15, Dean Drobot 4–5, frantic00 20–21, LD Productions 6–7, Progress man 12–13, Rawpixel.com 18–19, Sue Win 8–9.
All artwork, Brown Bear Books, Shutterstock and Square and Circus.
t-top, r-right, l-left, c-center, b-bottom

Websites
The website addresses in this book were valid at the time of going to press. However, it is possible that contents or addresses may change following publication of this book. No responsibility for any such changes can be accepted by the author or the publisher. Readers should be supervised when they access the Internet.

Words in **bold** appear in the Glossary on page 23.

Manufactured in the United States of America
CPSIA compliance information: Batch#AG/5660

CONTENTS

WHAT IS A SMARTPHONE?

The telephone was invented in 1876. We've come a long way since then! Early telephones were big. They were connected by wires. Modern smartphones are small. They have no wires. Even so, they can do much more. They are more like computers!

A PHONE IN YOUR POCKET

In the 1980s, people started to use cell phones. They could carry them around with them. But early cell phones were big and clunky. You could only use them for talking. Smartphones changed all that.

Cell phones don't need wires. They use invisible signals instead.

What Smartphones Do

Smartphones are a step up from basic cell phones. They do things that older phones couldn't.

You can still make voice calls.

You can also make video calls.

Users can send text messages. They can attach photos, movies, or sound files.

Cameras let you take photos and videos.

A smartphone can connect to the **internet**. You can browse the web.

Smartphones run **apps**. The apps help you with different tasks.

A smartphone can connect to other devices. It might connect to a watch or a speaker.

TOUCH SCREENS

Smartphones have big screens. These screens are special. They work by touch. You can tap or swipe. You can scroll and pinch. These gestures are instructions. They tell the phone what to do.

THAT'S ELECTRIC!

Did you know that you're electric? Your body can pass on electricity. Smartphone screens work by using this ability. They carry an electric charge. Some of it flows into your finger. But don't worry, it's not enough to hurt you!

No Gloves Allowed

You can't use a touch screen with gloves on. Their fabric can't pass on electricity. Only gloves made with special thread will work.

A light touch is enough on a smartphone screen. Pressing harder won't make a difference.

How It Works

The idea behind a touch screen is simple. But it takes complicated tech to make it work!

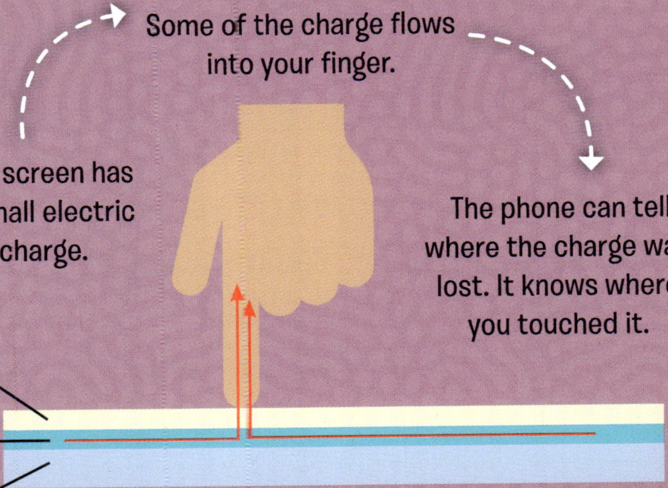

CELLULAR NETWORKS

Smartphones don't need wires. But they do need signals! These signals travel through the air. Tall towers send and receive them. These towers are called **masts**. You need to be near a mast to make calls. That's why there's a whole **network** of them.

RADIO WAVES

Energy can move from place to place. **Radio waves** are one way it travels. Smartphone signals travel as radio waves. So do TV signals. You can't see radio waves. You can't feel them, either. But they are everywhere!

A network has many cells. Each cell has its own mast.

Making a Call

Calling a friend seems quick and easy. But it takes a lot of steps.

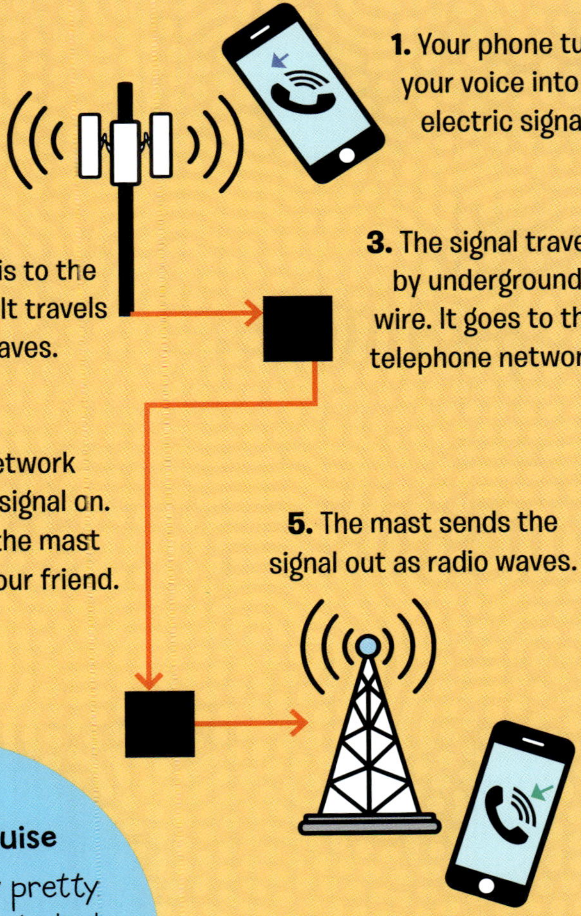

1. Your phone turns your voice into an electric signal.

2. It sends this to the nearest mast. It travels as radio waves.

3. The signal travels by underground wire. It goes to the telephone network.

4. The network passes the signal on. It goes to the mast closest to your friend.

5. The mast sends the signal out as radio waves.

6. Your friend's phone picks up the signal. It turns it back into your voice.

Masts in Disguise

Masts aren't very pretty. So some are made to look like other things. Masts can look like trees or flagpoles. Some even look like giant cactuses.

USING THE INTERNET

The internet connects people all over the world. Many people use computers to connect. But you can also use a smartphone. It uses the cellular network to connect. This uses radio waves. It's similar to the way voice calls work.

WI-FI

Smartphones can also connect using **Wi-Fi**. You don't need phone masts! You just need a Wi-Fi router. This is a small box. It sends out radio waves. Many homes have a Wi-Fi router. So do schools and offices.

A Wi-Fi router can't send signals very far. It only covers a small area.

The World in Your Pocket

A phone that connects to the internet is really useful. You can do so many things!

- Send and receive emails
- Stream music, movies, and TV shows
- Check the news, weather, and sports scores
- Shop online
- Make video calls
- Post on social media
- Chat using messaging apps
- Play online games

MAPS AND GPS

Most smartphones come with a map app. It shows maps for anywhere in the world. The maps show roads and rivers. They show shops and restaurants. It's like having a big atlas in your pocket.

TAKING IT FURTHER

Do you want to go somewhere? The map app will give you directions. It looks for traffic and other delays. It shows you the fastest route. The app shows your location as you move. This might seem like magic. But it's science! It works using **satellites** in space.

A map app tells you how long a journey will take. It estimates when you will arrive.

Finding Your Way

Smartphones use a system called GPS. It stands for Global Positioning System.

GPS uses satellites high above the Earth. There are about 30 of them.

The satellites send out signals. Your phone receives them.

The signals show how far away a satellite is.

You are always in range of at least four satellites.

Your phone needs three signals for GPS to work.

It sees how far away each satellite is. Then it uses math to figure out your location.

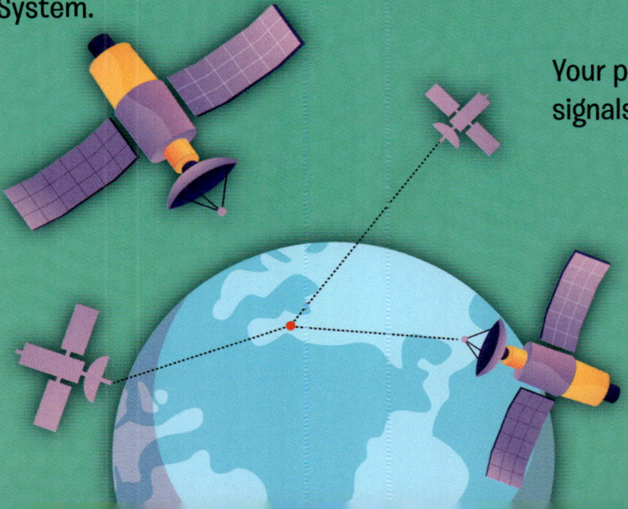

SMARTPHONE CAMERAS

Time to take a photo! Nearly all smartphones have cameras. They let you take photos. You can also shoot videos. There are usually two or more cameras on the back. There is another on the front. It is for taking selfies.

The screen lets you see what you're shooting. It's easy to get a good selfie.

GETTING THE RIGHT SHOT

Cameras can be complicated. But a smartphone does most of the work. It chooses the best settings. It also has **software** for editing your shots. This helps make them look good. You can share your photos with friends.

Multiple Cameras

Your phone chooses which camera to use. This means you always get the best shot.

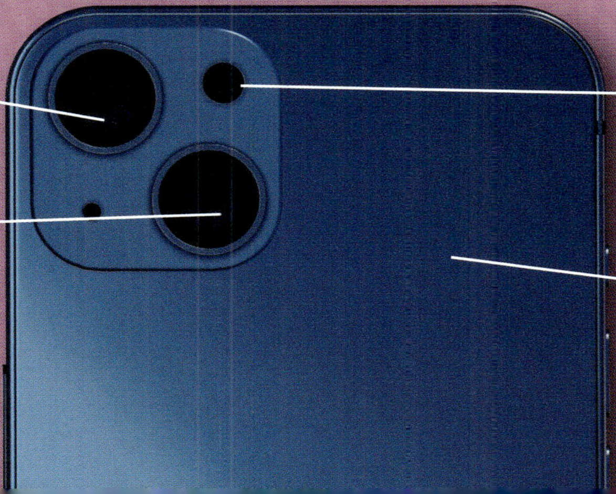

MAKING CONNECTIONS

Smartphones can do a lot. But they can't do it all! Sometimes they need help to do a job. They connect to other devices, such as a speaker. This is called "pairing." It uses a system called **bluetooth**. Devices send and receive **wireless** signals.

STAYING CLOSE

Bluetooth signals only travel about 30 feet (9 m). When two devices are close enough, they connect. They can share data. A smartphone can control headphones or a car stereo.

Bluetooth headphones have no wires to get tangled up.

Blast from the Past

Bluetooth is named for a Viking chief. His name was Harald Bluetooth. He got enemy tribes to cooperate. Bluetooth signals make devices cooperate!

Get Connected!

The bluetooth in a phone can connect to many different things.

You can play music through headphones.

Share files by connecting to a laptop.

You can play music for your friends on a speaker.

Connecting a keyboard lets you type more easily.

A smartphone can connect to your car's sound system.

A fitness tracker sends stats to the app on your phone.

GAMES AND MUSIC

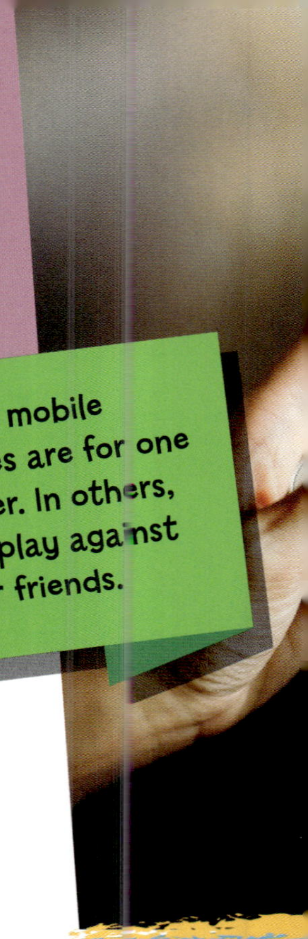

Some mobile games are for one player. In others, you play against your friends.

Smartphones help us stay connected. They make our daily lives easier. They can also keep us entertained! Many people play games on their phones. They listen to music or podcasts. You can even watch movies or TV shows. This is called **streaming**.

MOBILE GAMES

Many mobile games are free to play. The rules are easy to learn. They only take a short time to play. But you can play them over and over! Puzzle games are popular. So are adventure games. These games can pass the time on a bus ride.

Music Streaming

More than 600 million people use music streaming.

Many users pay a monthly fee. They can listen as much as they want.

Some streaming services are free. But they play ads.

The artist gets paid every time someone listens.

You can make a playlist of your favorite tracks.

You don't have to store tracks on your phone.

The app suggests other artists you might like.

A WORLD OF APPS

An app can help you arrange a vacation. Another app can translate menus.

Smartphones can do a lot. That's because they run apps. An app is a piece of software. It lets your smartphone do a job. An app can be simple, like a calculator. Other apps are more complicated.

WHAT TO CHOOSE?

There are thousands of different apps. You choose which ones to install. Some are free. Others are paid for. You download an app to your phone. Its **icon** appears on the screen. You tap it to use the app.

Apps at Your Fingertips

A phone can hold dozens of apps. There are many different types.

Time to Go!

An app's software is stored on your phone. Having too many can clog up your phone. Are there apps you don't use any more? Delete them to save space.

QUIZ

How much have you learned about smartphones? It's time to test your knowledge!

1. Which of these is the only one a smartphone cannot do?

a. make voice and video calls
b. connect to the internet
c. cook dinner

2. Why won't a touch screen work if you're wearing gloves?

a. the fabric blocks the transfer of electricity
b. you're not able to press hard enough
c. the gloves make your fingertips too big to select an icon

3. What does GPS stand for?

a. Geographic Pinpointing Satellite
b. Global Positioning System
c. Giant Pointy Stylus

4. What is the name of the system that lets devices connect wirelessly to each other?

a. greenleg
b. bluetooth
c. yellowmouth

The answers are on page 24.

GLOSSARY

app a computer program designed to do a particular job, usually on a smartphone or tablet

bluetooth a system that lets devices connect wirelessly to one another

icon a symbol that represents an app on a screen

internet a huge computer network that connects computers all over the world

mast a tall tower with equipment for sending and receiving phone signals

network a group of connected items, such as a system of cellular masts that send signals to each other

radio waves invisible waves of energy that travel through the air and can be used to send information

satellite an artificial object sent into orbit around Earth, which can send and receive communication signals

software a set of instructions that a computer follows in order to complete a task

streaming using an internet connection to listen to music or watch videos

Wi-Fi a system for connecting wirelessly to the internet, using a signal sent by a router

wireless a way of connecting to a network using radio waves instead of wires

FIND OUT MORE

Books

Cell Phones and Smartphones: A Graphic History. Blake Hoena, Graphic Universe, 2021.

Smartphones. Martin Gitlin, Cherry Lake Publishing, 2019.

Smartphones. Tammy Gagne, Cody Koala, 2018.

Websites

kids.britannica.com/students/article/smartphone/570926

spaceplace.nasa.gov/gps-pizza/en/

www.wonderopolis.org/wonder/how-does-a-phone-work

INDEX

Answers: 1. c; 2. a; 3. b; 4. b